Laís Fernanda Fausto de Oliveira
Carla Qualio Otero
Nathália Pavan Vianna

Environmental Tobacco Pollution (ETS)

Laís Fernanda Fausto de Oliveira
Carla Qualio Otero
Nathália Pavan Vianna

Environmental Tobacco Pollution (ETS)

And its consequences on the respiratory system and weight gain in Wistar rats

ScienciaScripts

Imprint

Any brand names and product names mentioned in this book are subject to trademark, brand or patent protection and are trademarks or registered trademarks of their respective holders. The use of brand names, product names, common names, trade names, product descriptions etc. even without a particular marking in this work is in no way to be construed to mean that such names may be regarded as unrestricted in respect of trademark and brand protection legislation and could thus be used by anyone.

Cover image: www.ingimage.com

This book is a translation from the original published under ISBN 978-613-9-73146-6.

Publisher:
Sciencia Scripts
is a trademark of
Dodo Books Indian Ocean Ltd. and OmniScriptum S.R.L publishing group

120 High Road, East Finchley, London, N2 9ED, United Kingdom
Str. Armeneasca 28/1, office 1, Chisinau MD-2012, Republic of Moldova, Europe
Printed at: see last page
ISBN: 978-620-7-76088-6

1

DEDICATORY

We would like to dedicate
this work to everyone who
contributed in any way.

ACKNOWLEDGEMENTS

We would firstly like to thank God for always guiding our paths and our parents for all the encouragement and support they have given us.

To our supervisor Ms Ana Rosa Crisci for all her support and dedication, and to our Masters who provided us with the learning we needed to complete our degree.

To the veterinarian Dr Marcelo Pardi Castro, the animal house technician Ramiro Nunes da Silva and the student Alan Carlos Fernandes, who helped us throughout our experiment.

"Early addictions are the strongest. It is through the new generations, and with them, that we will win the battle against smoking. " (Dr Hiroshi Nakajima, 2014).

SUMMARY

Environmental Tobacco Pollution (ETS) is the pollution generated by burning tobacco products indoors, making it one of the most important public health problems. This research is based on an experimental study, with the aim of analysing the morphological changes that occurred in the respiratory system and evaluating the changes in water and food consumption and body mass gain in adult Wistar rats subjected to exposure to cigarette smoke. The animals were divided into two groups, one of which was treated with smoke exposure and the other a control group subjected to the same regime without smoke. The histopathological results showed the presence of inflammatory processes, loss of cilia and thickening of the pulmonary alveoli. In relation to body mass, water and food consumption, there was a reduction in body mass and higher water and food consumption. This damage highlights the importance of understanding the effects of PTA on both the environment and human health.

Keywords: Respiratory Mucosa. Tobacco smoke pollution. Wistar rats.

SUMMARY

1 INTRODUCTION

Environmental Tobacco Pollution (ETS) is the pollution generated by burning tobacco products indoors, also known as Environmental Tobacco Smoke (ETS), which is generated by the sum of the smoke released into the environment, known as the secondary current, plus the smoke exhaled by the smoker (ANVISA, 2009).

The secondary stream has higher quantities of carcinogenic substances because they do not pass through the filter and occur at a lower temperature, so the combustion of the substances is incomplete (NUNES et al., 2011).

When a cigarette is lit, only part of the smoke is inhaled, but around 2/3 of it is released into the environment by the lit end of the product. The same happens with similar products, such as cigars, cigarillos and straw cigarettes (ANVISA, 2009).

When Christopher Columbus arrived in Central America, he observed that the indigenous people burned the leaves of a plant to smoke, with the aim of purifying, contemplating, protecting and strengthening the brave warriors, as well as believing that it had the power to predict the future. Over time, the seeds of this plant were taken to Europe by French ambassador Jean Nicot - the name that gave rise to the botanical term nicotine - and used as a medicine for the nervous system, becoming the most widely consumed drug in the world, tobacco (FERREIRA, 2002).

The first scientific study on environmental pollution was published in the USA in 1993 by the Environmental Protection Agency, which reported that passive exposure to cigarette smoke is harmful. In Brazil, the National Cancer Institute (INCA) carried out a study in 2008 showing that this exposure causes an average of seven deaths a day (INCA, 2009).

One of the biggest challenges facing the World Health Organisation (WHO) is smoking in the world. Today, smoking causes more deaths than other drugs, traffic accidents, suicides and fires, and it is estimated that approximately five million people die each year. Nicotine addiction is included in the International Classification of Diseases (ICD 10), as it is considered a chronic disease (FERREIRA, 2002).

According to Damé, Cesar and Silva (2011), if tobacco consumption remains constant until 2030, it could cause up to 8 million deaths a year worldwide, 80 per cent of which will occur in low-income countries.

According to data from the International Labour Organisation (ILO), pollution caused by tobacco smoke in the workplace causes at least 200,000 deaths every year (NUNES et al. 2011).

Nicotine addiction is compared to the effects of cocaine, alcohol and heroin, which are promoted by biopsychosocial processes, in which an individual who smokes 20 cigarettes a day receives more than 70,000 brain hits of nicotine a year. It reaches the brain in less than 10 seconds, releasing dopamine, endorphins and other hormones that are responsible for feelings of pleasure, concentration, humour and reducing withdrawal symptoms (lack of tobacco) (ZANIN; SANTOS, 2006).

One of the first laws to ban smoking in enclosed public or private spaces for collective use was Federal Law 9294/96, but it was not very effective due to the lack of parameters for enforcement. Over time, some Brazilian states realised the harm that environmental tobacco pollution caused to the population and the importance of having a smoke-free environment, and with the improvement of this law, specific environments were created for smokers, the so-called smoking rooms (CÉZAR, et al., 2014).

In view of the concern and work carried out by government bodies, Federal Law No. 12,546 came into force in 2011, banning smoking in closed environments for collective use, whether private or public. The body responsible for overseeing this law is the National Health Surveillance Agency (ANVISA) (SILVA et al., 2014).

According to the Global Smoke Free Partnership (2008) apud Cézar et al. (2014), say that:

> "There are countries with national laws where there are no exemptions for smoke-free environments, or with exemptions limited to residential or semi-residential spaces. Countries in this category do not allow smoking rooms [...] but in some countries smoking rooms are allowed and there are some exemptions that apply to a small number of places, such as smoking rooms."

Brazil has already been taking various measures to reduce smoking, for example with regard to advertising. According to Silva et al (2014, p. 541):

> Brazil stands out as the first country to ban descriptors on packaging, the second to insert warnings with phrases and images on packets and one of the few to restrict advertising and to impose a veto on the food industry in relation to the commercialisation of products that simulate tobacco derivatives as well as their packaging.

The country's attitude of using warning adverts instead of encouraging consumption has proved to be effective, according to a study carried out by the Datafolha Institute (2012), cited by Silva et al (2014, p. 541), "[...] 67% of smokers said they felt like giving up smoking".

The harm caused by tobacco has been known for some time, but the scientific community's concern about passive smoking is recent. Passive smoking is considered the third leading cause of preventable death in the world, second only to active smoking and excessive alcohol consumption (COELHO; ROCHA; JONG, 2012).

Passive smokers are defined as people who live in enclosed spaces with smokers but do not smoke. They are exposed to the components present in environmental cigarette smoke, such as nicotine, carbon monoxide and other substances, which are toxic and carcinogenic, and have a 30 per cent greater chance of developing lung cancer than people who are not exposed (FERREIRA, 2002).

Newborns who come into contact with tobacco during pregnancy and/or breastfeeding can also be considered passive smokers. It is estimated that around 700 million children suffer from this exposure (COELHO; ROCHA; JONG, 2012).

This exposure to PTA is proven by measuring a product of nicotine decomposition, cotinine, the most widely used marker, found in the blood and urine of non-smokers in high concentration and for longer. Nicotine absorbed by passive smokers does not lead to dependence, as it takes longer to be metabolised and is excreted quickly, in around 30 to 110 minutes, while cotinine has a half-life 10 times longer, lasting from 19 to 40 hours (MELLO et al., 2005).

Cotinine accumulates in the foetal circulation, which can contribute to and induce premature labour and miscarriage in smokers, as well as the fact that this exposure is more significant during breastfeeding. This is proven by the levels of this substance found in the children of smoking mothers, which are equivalent to active smokers (MELLO; PINTO; BOTELHO, 2001).

The National Household Sample Survey (PNAD) carried out in 2008 estimated for the first time the degree of exposure to ETS in the Brazilian urban adult population, proving that individuals who live with a smoker inhale 1% of the smoke emitted by 20 cigarettes during a day (PASSOS; GIATTI; BARRETO, 2011).

According to studies carried out by ANVISA (2009, p.5):

> Environmental tobacco smoke contains practically the same composition as smoke inhaled by smokers: around 4,000 compounds, of which more than 200 are toxic and around 40 are carcinogenic. However, the levels of these contaminants are higher, with an average of 3 times more nicotine, 3 times more carbon monoxide and up to 50 times more carcinogenic substances being found than the smoke inhaled by the smoker. This is because the smoke that comes out of the lit tip is not filtered.

A major problem generated by tobacco is the deforestation of large forests to harvest the raw material. Wood-fired ovens and greenhouses are also used to dry the tobacco leaves before they go to the industry. For each tree burnt, 300 cigarettes are produced, meaning that a smoker who consumes a packet a day kills a tree in less than 15 days (CÉZAR, et al., 2014).

After cigarettes are processed, around 150 chemical substances are added to them, such as pesticides, organic and metallic components. In addition, tobacco itself contains radioactive components such as lead (210 Pb) and polonium (210 Po). All of these substances are also present in its smoke and some carcinogens catalyse the transformation of aggressive agents by initiating the mutation of cells in the respiratory system (DUARTE et al, 2006).

Of the components present in cigarettes, 68 have been identified as carcinogenic, according to Zanin and Santos (2006, p.15):

- Ammonia - also used in products to disinfect bathrooms;
- Acetone - also used to remove nail polish and paints;
- Arsenic - an insecticide that is also poisonous to humans;
- Cyanide - poison used in gas chambers during the Second World War;
- Toluene - industrial solvent;
- Butane - used as lighter gas;
- Carbon monoxide - toxic gas emitted in car fumes;
- DDT - insecticide;
- Mothballs - a product we use to kill moths and cockroaches;
- Cadmium - used in car batteries.

Passive smokers suffer the immediate effects of Environmental Tobacco Pollution, such as: irritation of the eyes, nose, throat and lungs, coughing, headaches, increased allergic problems, an increase in the number of respiratory infections in children and a rise in blood pressure (COELHO; ROCHA; JONG, 2012).

In active and/or passive smokers, the histopathological alterations found are secretory hyperplasia and hypertrophy, which are the result of the continuous chronic aggression of tobacco in the airways. As a consequence, there is an increase in mucus production and a decrease in ciliary movements in the hair cells, causing an inflammatory process due to an increase in the number and size of goblet cells (SALLA et al., 2009; TAMASHIRO et al., 2009; DUARTE et al., 2006).

In view of the data presented, an experiment was carried out in which rats were subjected to cigarette smoke in order to assess the effects of nicotine and other substances on the respiratory system and changes in the animals' body mass and nutrition, since these data are still scarce in the literature.

Despite this, government bodies have already been creating measures to minimise these effects and research of this kind could provide biologists with information to help draw up and improve healthier public policies, reinforcing the role of this professional in protecting living beings and the environment.

2 MATERIALS AND METHODS

A review of the specialised literature was carried out between January and September 2014, which consisted of consulting books and periodicals in the Prof. Nicolau Dinamarco Spinelli Library at the Barão de Mauá University Centre. The scientific articles were selected by searching the Internet in scientifically credible databases such as SciELO, Medline, Google Scholar and the thesis banks of USP, UNESP and Unicamp.

The search for information sources in the databases was carried out using keywords and terminologies registered in the Health Sciences Descriptors (DeCS) created by the Virtual Health Library, based on the U.S. National Library of Medicine's Medical Headings, which allows the use of common terminology in Portuguese and English.

Articles were searched using descriptors in Portuguese and English.

The keywords and descriptors used in the search were: Smoking, Environmental Tobacco Pollution (ETS), smoker, nicotine, passive smoker, air pollution, cigarette smoke inhalation and respiratory epithelium.

The inclusion criteria for selecting the studies/articles were: selecting studies that reported on the problems generated by passive smoking for human beings and the environment, as well as studies on the influence of inhaling this smoke on weight gain, water and food consumption.

Specific studies reporting on the problem of smoking in small places such as schools, communities and neighbourhoods were excluded.

The material collected was selected in order of priority from the most recent to the oldest, covering the years 1997 to 2014.

This review sought to study and understand the main concepts about the harm caused by tobacco to the environment and the respiratory system of people who don't smoke but inhale this smoke, as well as the influence of inhaling this smoke on weight gain, water and food consumption.

This chapter describes the selection, the description of the groups assessed and the procedures carried out in this study.

This research was carried out at the Barão de Mauá University Centre, in Ribeirão Preto, from April to June 2014, with the aim of studying the morphological changes in the respiratory system of Wistar rats subjected to cigarette smoke.

According to Gil (2006), this study is an experimental investigation in which the object of study is the pathological alterations of the respiratory system under the action of cigarette smoke.

The observation of the effects that the variables produce on the object constitutes the research, in its cause-effect relationship (GIL, 2006). Considered the most prestigious design in science, in experimental research the researcher is an active agent.

The experiment followed the regulations proposed in Law No. 11,794 of 8 October 2008, which establishes procedures for the scientific use of animals, and was approved by the Ethics Committee for Research and Animal Experimentation (CEPAN) of the Barão de Mauá University Centre (protocol 210/2014).

2.1 Animals

F male Wistar rats weighing between 200g and 250g were used and divided into 2 groups: Group A: exposed daily to the smoke and Group B: subjected to the same regime inside the box, but without the smoke.

2.2 Cigarette smoke exposure protocol

Exposure to smoke followed a protocol according to Nolan et al. (1985), which was divided into two phases: the adaptation phase and the experimental phase.

The adaptation phase refers to the first 5 days in which the animals in group

A were exposed to cigarette smoke for a period of 1 hour/day.

The experimental phase began immediately after the adaptation phase. The animals in group A were exposed daily to the smoke from burning 4 cigarettes, one after the other, for 1 hour. Exposure took place daily for 56 days.

Group B was subjected to the same regime inside the box, but without the smoke. The exposures were standardised and carried out at approximately the same time of day.

2.3 Inhalation System

The inhalation system consisted of a polypropylene box measuring 45x28x18 cm, which was divided into 2 compartments by a perforated screen already made by a Biology student (fig. 1) and (fig. 2 p. 22).

Figure 1 Polypropylene box

Source: Personal archive.

Figure 2 Closed box.

Source: Personal archive.

The larger compartment was used for mice (up to 5 simultaneously) and the smaller one for cigarettes (fig. 3 p. 23). At the opposite end of the box, a fan (0.14 A computer cooler) was attached, connected to the 110 volt mains via an electrical transformer that supplied voltage from 3 to 12 volts. This generated a continuous flow of air inside the box: the smoke was sucked in by the fan, reached the rats and came out on the opposite side of the box. During the experiment, the box was closed with a wooden lid and lined with wood shavings (SILVA et al., 2011).

Figure 3 Box divided into two compartments

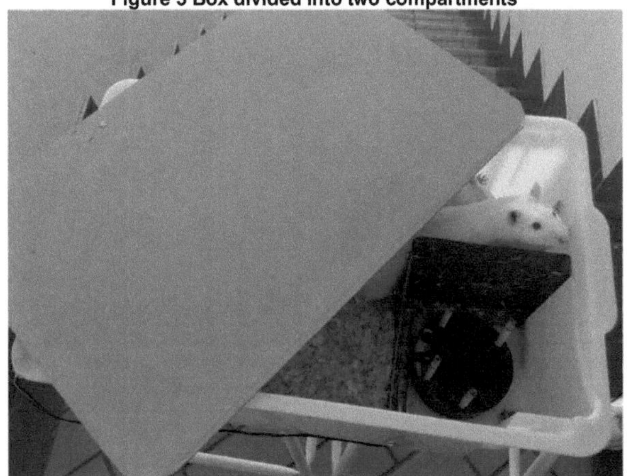

Source: Personal archive.

The brand of cigarette used was Marlboro (fig. 4) (Phillip Moris), red packet with 20 units, each cigarette containing 0.8 mg of nicotine, 10 mg of tar and 10 mg of carbon monoxide (MELLO et al., 2005).

Figure 4 Marlboro cigarette (Philip Moris).

Source: Personal archive.

2.4 Weighing the animals

The animals were weighed daily on a Gehaka BK 500 semi-analytical balance (fig. 5), and the data was stored on a special form.

Figure 5 Weighing the animals.

Source: Personal archive.

Daily food consumption was determined from the difference between the amount of feed offered to the animals and the amount discarded by the animal, always taking the previous day as a reference, on the same scales used to weigh the animals, and the data obtained was stored on a separate form.

Daily water consumption was estimated from the difference between the amount offered and the amount discarded. Both food and water intake were *ad libitum* by the animals, which were kept in individual cages (fig. 6 p. 25).

Figure 6 Individual cages.

Source: Personal archive.

The results were analysed using the mean and standard deviation, using the Student's *t-test to* compare the two groups. Despite the small sample size (n=8), it was possible to apply a parametric test (two-tailed unpaired Student's t-test) due to the fact that weight gain is actually an average of 57 weighings, which guarantees the normality of the data.

2.5 Statistical Planning/Experimental Design

The study is characterised as epidemiological, longitudinal, experimental, descriptive and analytical. The central tendencies of morphological and morphometric parameters, body mass, food and water consumption of the animals in the exposed and control groups will be compared using the non-parametric Mann-Mann test.

Whitney, due to structural properties of the data that constitute violations of the assumptions for carrying out the parametric t-Student mean comparison test, such as the fact that these data do not have a normal distribution.

This test does not make any assumptions about the sample size for each group compared, but the larger the size of the groups compared, the greater the power of the test to detect possible differences between them.

It is known that for samples greater than or equal to eight, i.e. when comparing groups made up of at least 8 individuals each, a good approximation to the normal distribution is obtained for the data, so that the test has power close to that of parametric tests, offering more accurate and precise results (CALLEGARI-JACQUES, 2003; ARANGO, 2009). Thus, 16 animals were requested, making two groups of 8 animals, one group of 8 to compare the morphological changes that occurred without smoke inhalation and the other group of 8 subjected to the action of smoke.

2.6 Laparotomy

After 56 days, the animals were euthanised in a CO_2 chamber.

After this process, a laparotomy was performed (fig.8 p.27) and the respiratory system organs were removed: lungs, trachea and larynx for histopathological analysis.

The lungs of each of the animals in the control and treated groups were isolated and weighed (fig. 8D p. 27) in order to compare the masses, according to the Mann-Whitne test. The relative mass of the organ was calculated using the formula:

Figure 7 Formula for the relative mass of the animal's organ.

$$\text{Massa relativa do órgão (\%)} = \frac{\text{massa do órgão (g)}}{\text{massa corporal (g)}} \times 100$$

Source: Dallegrave, 2003.

Figure 8 Laparotomy of the animal.

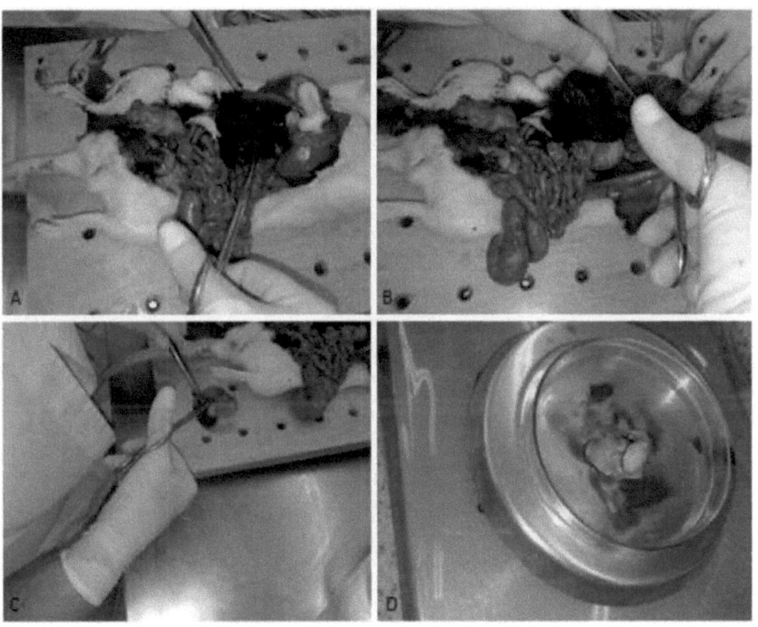

In A: Total exposure of internal organs. In B: detachment of the diaphragm, removal of the lung
In C: isolation of the lung; In D: weighing of the lung.
Source: Personal archive.

3 RESULTS

This chapter describes the results of the experiment, showing comparisons between the experimental and control groups in terms of body mass, food and lung mass and water consumption, as well as the histopathological results.

3.1 Average body mass, food and water consumption

The statistical analyses in Table 1 (p. 29) show the average body mass gain, food and water consumption of the animals in the experimental group (passive smokers) throughout the experiment (56 days). Table 2 (p. 30) shows the same items, but for the control group (non-smokers). These data were collected before the animals were placed in the inhalation box.

Table 1 Body mass gain (g), food consumption (g) and water consumption (ml).
Passive smoker

	Gain	Ration	Water
1	2,59	42,23	83,33
2	2,46	42,82	71,70
3	0,98	37,15	72,95
4	2,35	39,33	68,93
5	2,24	45,74	92,55
6	2,41	45,24	94,72
7	4,28	47,29	96,5
8	2,65	38,04	66,79
Media	2,49	42,23	80,93
Standard error of the mean	0,32	1,33	4,36

Source: Personal archive. Data collected from 22/04/2014 to 30/06/2014.

Table 2 Body mass gain (g), food consumption (g) and water consumption (ml).

Control			
	Gain	Ration	Water
17	4,64	39,86	65,80
18	4,45	38,97	65,09
19	5,43	42,01	68,39
20	2,91	36,85	61,79
21	3,61	35,42	51,61
22	2,72	40,50	81,61
23	3,72	43,05	86,48
24	1,92	33,38	63,48
Media	3,67	38,75	68,03
Standard error of the mean	0,40	1,17	3,94

Source: Personal archive. Data collected from 22/04/2014 to 30/06/2014

The results obtained in tables 1 and 2 are shown separately in the graphs below.

Graph 1 (p. 31) shows the average water consumption of the two groups over the 56 days of the experiment.

Graph 1 Water consumption.

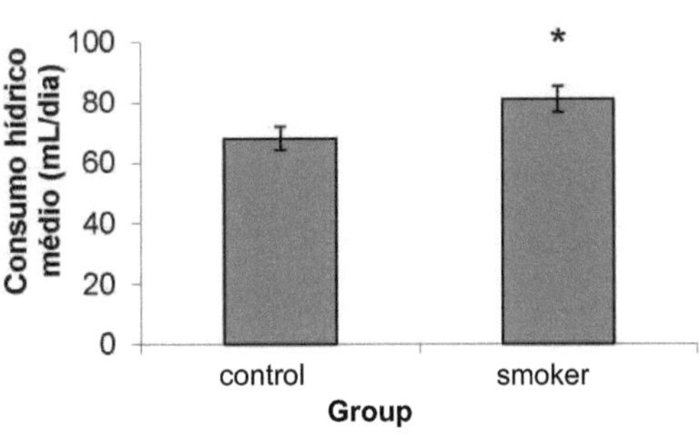

(*) Significant result.

Source: Personal archive. Data collected from 22/04/2014 to 30/06/2014.

Graph 2 shows the average food consumption of the experimental and control groups throughout the experiment.

Graph 2 Food consumption.

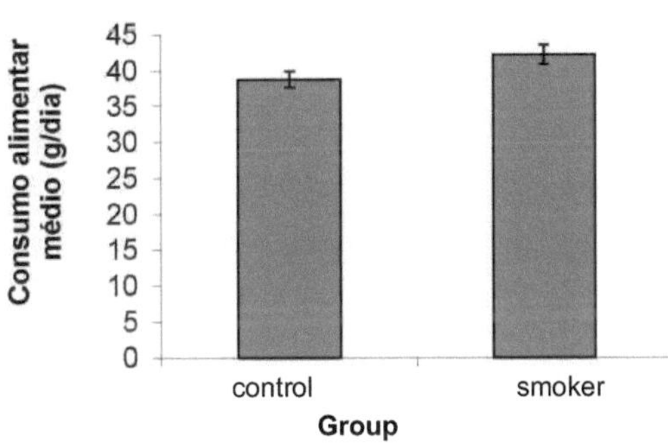

Source: Personal archive. Data collected from 22/04/2014 to 30/06/2014.

Graph 3 (p. 32) shows the difference in the average body mass acquired by the two groups over the course of the experiment.

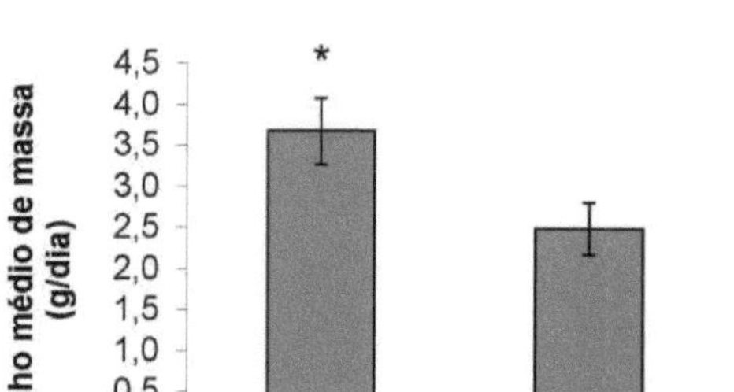

Graph 3 Body mass.

(*) Significant result.
Source: Personal archive. Data collected from 22/04/2014 to 30/06/2014.

3.2 Histopathological results

In the photomicrographs of the histopathological aspects of the larynx (fig. 9 p. 33), it was observed in A (control group) that the larynx had a normal appearance, with preservation of the epithelium (short arrow) and underlying tissue (connective and cartilage) (white arrow). In D: the smoker group, the larynx showed discontinuous epithelial areas with normal epithelium (short arrow) and a marked inflammatory process in the underlying subcutaneous tissue (white arrow).

In the photomicrographs of the histopathological aspects of the trachea (fig. 9 p. 33), it was observed in B (control group) that the trachea had a normal-looking architecture, with preservation of the ciliated epithelium (short arrow), intact subcutaneous tissue with tracheal glands (white arrow). In E: smoker group, the trachea showed a marked inflammatory process in the connective tissue (white arrow).

In the photomicrographs of the histopathological aspects of the lung (fig. 9 p. 33) it was observed in C (control group) that the lung parenchyma was totally preserved, bronchioles and interalveolar septa were normal (dotted arrow), alveolar lumen and bronchioles without vascular congestion. In F: the smoker group, the lung parenchyma was disorganised, with pulmonary congestion and thickening of the alveolar septa due to vascular congestion (dotted arrow).

Figure 9 Histopathological results.

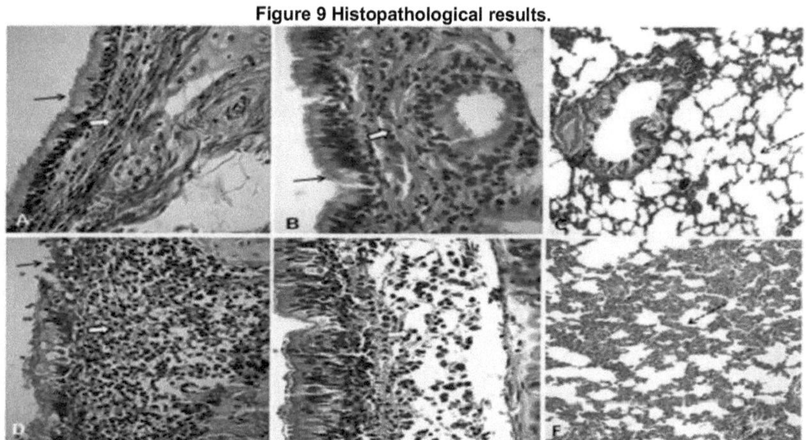

Photomicrographs of the histopathological aspects of the larynx, trachea and lungs of rats, respectively. In A, B and C control group. In D, E and F the smoking group. H.E. staining. Final magnification: 200X.
Source: Personal archive.

4 RESULTS

In this chapter, in order to fulfil the previously proposed objective of studying the morphological alterations and changes in body mass and nutrition that environmental tobacco pollution causes in Wistar rats, a critical analysis of the results obtained was carried out.

The rat was chosen as the experimental model because it is easy to maintain and handle, and because the particularities of this animal are well known, it is possible to recognise the morphological characteristics of the organs, which are very similar to those of humans.

Tobacco pollution causes problems not only for the environment, but also for living beings, and the lack of information on the subject is one of the obstacles to achieving the desired change in attitude towards the harm it causes. As such, one of the objectives of the biological sciences graduate is to inform students about these harms at school. According to Salla et al. (2009), the school is the basis for promoting health education by providing information to raise awareness about the consequences of PTA, to understand the problems generated in biological systems as a result of passive smoking and to develop political-pedagogical projects to prevent diseases related to this pollution.

With the results obtained in this study, it was possible to observe that the adult animals subjected to cigarette smoke exposure for 56 days showed a reduction in body weight when compared to the control group. These results are in line with those obtained by Silva (2009), who under very similar conditions, but in young rats during 30 days of passive inhalation, obtained a significant reduction in the animals' weight. In this study, there was no reduction in food intake in the experimental group, probably because the time of exposure to smoke may not have been sufficient for nicotine to promote such an appetite suppressant reaction, accelerating metabolism.

These results can be compared with those obtained by Randi et al. (2008), who state that the action of nicotine causes alterations in the digestive system and this makes it difficult to absorb food, as a consequence increasing food and water consumption. In their work, the rats were exposed to cigarette smoke every day for an hour for 7 days over a period of 6 months (chronic smokers).

In the study by Gonçalves-Silva, Lemos-Santos and Botelho (1997), they observed that animals exposed to smoke from 2 cigarettes a day, twice a day, for 30 days, had weight loss, but water consumption showed no statistically significant difference.

The results obtained by Chatkin and Chatkin (2007) in a review on smoking and body weight, state that passive smokers when exposed to cigarette smoke, in contact with nicotine, cause an increase in the number of hormones such as dopamine and serotonin, which are neurotransmitters that inhibit food intake.

These same authors also state that all these data relating body weight to the action of nicotine are controversial, as they depend on the sex of the individual, whether they are physically active or sedentary, whether they quit smoking and returned to the habit later, the type of food eaten after quitting, genetic variations such as mutations in the gene that regulates leptin activity, so this is a promising avenue for further research.

In addition to these data, this study sought to analyse the organs of the respiratory system from a morphological point of view, observing a marked inflammatory process in the underlying subcutaneous and connective tissue, loss of cilia and a thickening of the pulmonary alveoli. These alterations in ciliogenesis are in agreement with Tamashiro et al. (2009), who, using a culture of respiratory epithelium cells obtained from the nasal septum of mice, demonstrated that exposure of the respiratory epithelium to both the particulate and gaseous phases of cigarette smoke causes a significant reduction in the percentage of cilia development.

Also according to studies by Tamashiro et al. (2009), depending on the concentration and time of exposure to cigarette smoke, it can cause hyperplasia and metaplasia with keratinisation, as well as thickening and inflammation of the submucosa, caused by defence cells of the immune system.

5 CONCLUSION

The purpose of this chapter is to present the conclusion of the research carried out.

The results obtained using this experimental model confirm that the methodology applied was able to demonstrate the histopathological changes in the treated group and the preservation of the normal characteristics of the control group, as well as changes in food and water consumption and the body weight of the animals in the experimental group.

Exposure to cigarette smoke during the stipulated periods caused: a reduction in body mass, higher water and food consumption; histopathological changes in the respiratory system, such as: inflammatory processes in the tissues, loss of cilia, and thickening of the pulmonary alveoli.

In this research, in addition to the histopathological results, it was observed that the time of exposure to Environmental Tobacco Pollution not only affects the respiratory system, but also involves the other systems and the metabolism of the individuals in contact with it.

Biologists are intrinsically concerned about the environment and living beings, so they are also responsible for studying, discussing and acting to demonstrate the damage caused by this pollution, as was done in this research by carrying out demonstrative experiments.

It is therefore important to emphasise the need for research and discussions involving the PTA, with the aim of providing additional information about it.

REFERENCES

ANVISA, ANVISA in the reduction of involuntary exposure to tobacco smoke. **National Health Surveillance Agency**, p. 24. 2009.

CALEGARI-JACQUES; SIDIA, M. **Biostatistics: principles and applications.** 1 ed. Porto Alegre: Artmed, 2003.

CÉZAR, A.; ELVIS, D.; SANTOS, J. A.; BARBOSA, M. A.; NASCIMENTOS, R. A. B.; SOUZA,T.; ARAUJO, P. J. P. The problems caused by active smokers to passive smokers at UNIT and the creation of a smoking area. **Cadernos de Graduação - Ciências Exatas e Tecnológicas Unit**. Aracaju. v. 2, n. 1, p. 11-20, mar, 2014.

CHATKIN, R.; CHATKIN, J. M. Smoking and weight variation: can pathophysiology and genetics explain this association? **J. Bras. Pneumol,** Porto Alegre. v. 33, n. 6, p. 712-719, 2007.

COELHO, S. A.; ROCHA, S. A.; JONG, A. C. Consequences of passive smoking in children. **Rev. Cienc. Cuid. Saúde.** São Paulo. v. 11, n. 2, p. 294-301, Apr-Jun. 2012.

DALLEGRAVE, E. **Reproductive toxicity of the glyphosate herbicide Roundup® in Wistar rats**. 2003. 200 f. Thesis (Doctorate in Veterinary Sciences) - Veterinary Faculty, Federal University of Rio Grande do Sul, Porto Alegre.

DAMÉ, J. L. D.; CESAR, J. A.; SILVA, S. M. Temporal trend of smoking in urban population: a population-based study in Southern Brazil. **Cad. Saúde Pública**, Rio de Janeiro, v. 27, n. 11, p. 2166-2174, nov. 2011.

DUARTE, J. L.; FARIA, F. A. C.; CEOLIN, D. S.; CESTARI, T. M.; ASSIS, G. S. Effects of passive cigarette smoke inhalation on the vocal folds of rats. **Rev. Bras. Otorrinolaringol**, São Paulo. v. 72, n. 2, p. 210-216, mar-abr. 2006.

FERREIRA, A. M. **Tabagismo**, 2002. 33 p. Monograph (Degree in Biological Sciences) - Faculty of Health Sciences, Brasília University Centre, Brasília.

GIL, A. C. **Como elaborar projetos de pesquisa**. 4 ed. São Paulo: Atlas, 2006. p. 47-49.

INCA. **The "myth" (?) of passive smoking.** Available at: <http://www1.inca.gov.br/tabagismo/atualidades/ver.asp?id=1208>. Accessed on: 17 March 2014.

Anti-smoking law
Available at
<http://www.leiantifumo.sp.gov.br/portal.php/lei>. Accessed on: 23 March 2014.

MELLO, P. R. B.; OKAY, T. S.; DORES, E. F. G. C.; BOTELHO, C. Markers of smoking exposure in lactating rats using a passive exposure model from early gestation. **Pulmão,** Rio de Janeiro, v. 14, n. 4, p. 289-293, 2005.

MELLO, P. R .B.; PINTO, B. R.; BOTELHO, C. Influence of smoking on fertility, pregnancy and lactation. **J. Pediatr,** Rio de Janeiro. v. 77, n. 4, p. 257-264, 2011.

NAKAJIMA, H. **WHO campaigns against cigarettes.** Available at: < http://www1.folha.uol.com.br/fsp/1995M/30/cotidiano/24.html >. Accessed on: 14 October 2014.

NOLAN, J.; JAMES M. D.; JENKINS; ROGER A. Ph. D.; KURIHARA; KUNIHIRO M. D.; SCHULTZ; RICHARD C. M. D. The acute effects of cigarette smoke exposure on experimental skin flaps. **Plast Reconstr. Surg.** v. 75, n. 4, p. 544-551, 1985.

NUNES, S. O. V.; CASTRO, M. R. P.; LANSSONI, M. M. B. S.; MACHADO, R. C. R. Tobacco-Free Environment. In: NUNES, S.O.V., CASTRO, M.R.P., **Tabagismo:** abordagem, prevenção e tratamento. Londrina: EDUEL, 2011. chap. 3, p. 57-64.

MONTEIRO, R.; BRANDAU, R.; GOMES, W. J.; BRAILE, D. M. Trends in animal experimentation. **Rev Bras Cir Cardiovasc,** São Paulo. v. 24, n. 4, p. 506-513, 2009.

PASSOS, V. M. A.; GIATTI, L.; BARRETO, S. M. Passive smoking in Brazil: Results of the Special Smoking Survey, 2008. **Ciência & Saúde Coletiva.** Minas Gerais, v. 16, n. 7, p. 3671-3678, 2011.

RANDI, B. A.; STEFAN, L. F. B.; ZANCHETTA, C. M. C.; CUNHA, M. R.; CALDEIRA, E. J. Body weight variation in rats submitted to chronic passive smoking. Perspectivas Médicas, **Red de Revistas Científicas de América Latina, el Caribe, Espana y Portugal,** São Paulo. v. 19, n. 2, p. 5-8, jul.-dez., 2008.

SALLA, L. F.; SALLA, R. F.; FIGUEIRA, A. C. M.; MONTEDO, L.; ROCHA, J. B. T. "Pulmão e sua turma": The effects of environmental tobacco pollution on the respiratory epithelium. An experience from the perspective of *empowerment education* in health promotion at school. **National Meeting of Research in Science Education**, Florianópolis. 2009. 14 p.

SILVA, J. B.; GAZZALLE, A.; MANO, L. F. M.; COCOLICHIO, F.; ZAMPIERI, T. J.; PELLIZZARI, A. C. Is it possible to statistically validate a low-cost experimental device with 10 rats for research into passive smoking? **Rev. Bras. Cir. Plast.** v. 26, n. 2, p. 194-197, 2011.

SILVA, P. E. **Effect of passive smoking and associated physical exercise on glucose transporter glut4 expression in rat muscles.** 2009. 92 f. Dissertation (Master's Degree in Physiotherapy) - Faculty of Science and Technology, Universidade Estadual Paulista, Presidente Prudente.

GONÇALVES-SILVA, R. M. V., LEMOS-SANTOS, M. G., BOTELHO, C. Influence of smoking on weight gain, body growth, food and water consumption in rats. **J Pneumol** v. 23, n. 3, p. 124-130, May-June 1997.

SILVA, S. T.; MARTINS, M. C.; FARIA, F. R.; COTTA, R. R. M. Combating smoking in Brazil: the strategic importance of government actions. **Ciência & Saúde Coletiva**, v. 19, n. 2, p. 539-552, 2014.

TAMASHIRO, E.; COHEN, N. A.; PALMER, J. N.; LIMA, W. T. A. Effects of cigarette smoking on the respiratory epithelium and its role in chronic rhinosinusitis. **Braz J Otorhinolaryngol**, v. 75, n. 6, p. 903-907, 2009.

TAMASHIRO, E.; XIONG, G.; ANSELMO-LIMA, W. T.; KREINDLER, J. L.; PALMER, J. N.; COHEN, N. A. Cigarette smoke exposure impairs respiratory epithelial ciliogenesis **American Journal of Rhinology & Allergy,** v. 23, n. 2, p. 117-122, 2009.

ZANIN, I; SANTOS, V. **Analysis of the "Cognitive Behavioural Approach to Smoking Control and Treatment" programme in the municipality of Caçador - SC.** 2006. 79 p. Monograph (Graduation in Nursing) - University of Contestado, Santa Catarina.

APPENDIX

CENTRO UNIVERSITÁRIO "BARÃO DE MAUÁ"
COMITÊ DE ÉTICA EM PESQUISA E EXPERIMENTAÇÃO
ANIMAL – CEPan

CERTIFICADO

Certificamos que o Protocolo registrado no CEPan sob o nº **210/2014**, intitulado :
**"Poluição Tabagista Ambiental (PTA) e suas Consequências nos Sistemas Biológicos.
Estudo Experimental em Ratos.",** sob a responsabilidade do Prof.Drª. Ana Rosa
Crisci, está de acordo com a Lei 11.794 de 08/10/2008 e foi considerado
APROVADO pelo *Conselho do Comitê de Ética em Pesquisa e Experimentação
Animal-CEPan*, do Centro Universitário Barão de Mauá na presente data.

Ribeirão Preto, 1 de Abril de 2014

*Profª Drª Eneida P. dos S. de Aguiar
Presidente do CEPan-B.M. e
Conselheira do COMDEA-
Prefeitura Municipal de
Ribeirão Preto - PMRP*

Rua Ramos de Azevedo 423, Jd Paulista – Ribeirão Preto – SP Cep 14090-180 Fone 3603-6624-
Fax 3603-1050 e-mail: cepan@baraodemaua.br

Printed by Books on Demand GmbH, Norderstedt / Germany